How to project manage your self-build home

- Learn how to build your house as a construction project manager
- Learn and adopt easily understood practical tips
- Learn how to create a sustainable budget
- Learn how to create a reliable time schedule
- Learn the correlation between quality and cost
- Learn the correlation between actions & design
- Ready templates to use in MS Word & Excel provided

Niclas Kingston MBA, PMP, PMI-RMP & PMI-ACP

Copyright © Niclas Kingston
First edition
All rights reserved.

No part of this publication may be reproduced, stored in a retrieval system or transmitted in any form or by any means, electronic, mechanical, photocopying, recording, scanning or otherwise, except under terms of the copyright, designs and patents act 1988 or under the terms of a license issued by the copyright licensing agency ltd.

Legal disclaimer

The author makes no representations or warranties with respect to the accuracy or completeness of the contents of this work and specifically disclaim all warranties, including without limitation warranties for a particular purpose. no warranty maybe created or extended by sales or promotional materials. the advice and strategies contained herein may not be suitable for every situation. the author shall not be liable for damages arising here from. the fact that an organization or website is referred to in this work as a citation and/or a potential source of further information does not mean that the author or the publisher endorses the information the organization or website it may provide or recommendations it may make.

This work is dedicated to my two sons, Henry and Theodor, whose understanding and support have been much appreciated.

Content

INTRODUCTION	11
An Idea/Vision	14
Inspiration	15
PLANNING	16
ASSUMPTIONS & EXPECTATIONS	19
FEASIBILITY ANALYSIS	23
PLANNING PHASE – PLAN, PLAN AND PLAN	26
Break the Project Down	28
WBS – WORK BREAKDOWN STRUCTURE	30
Activity = Work Package	30
The Organization – The Team	36
Choice of Production Method / Finish	39
Prefabricated Products	41
Think Things Through – Avoid Implications	42
When to Start Production	46

Visual Details	46
Understanding Correlation between Activities	48
People Involved	50
Family	51
Contractor and Sub-contractor	52
References – the Contractor	54
References – the Consultants	57
Environmental aspects	57
Temporary Living	58
Drawings	59

EARLY PLANNING DONE — 64

The Project/House itself	65
Building Site	66
Tidy Project	68
Temporary Solution	69
Temporary Building Actions	69
Building Material Outside	70
Foundation of the Building	71

Disability	74
Windows	75
The Sun	76
Windowsill	76
Acoustics	77
HI-FI System	78
Fire Precautions	78
Motion Detector	79
Characteristics of the walls inside the house	79
Garage	79
Wet rooms	80
Storage	80
Fireplace	81
Security	81
Toilet	82
Garden	82
Driveway	83
Miscellaneous	83

Preparations for Future Needs   84

SWOT ANALYSIS   85

SWOT Actions   85

Opportunities   87

RISKS   89

Actions   90

Risk Assessment: 90

BUDGET   92

TIME SCHEDULE   96

PLANNING DONE   101

THE AGREEMENT   102

The Agreement with the Contractor   102

Fixed Price   102

Time and Materials   104

Statement of Work (SOW)   104

Room description   105

Specifications and Measurement   **107**

EXAMPLE OF A CONTRACT   **107**

The Contract   **108**

To-do List   **109**

Safety   **109**

Production   **111**

Independent Surveyor   **112**

Maintenance   **113**

Accommodation for Tenant   **113**

SUMMARY   **115**

FINAL THOUGHTS   **117**

WORDLIST   **118**

Acknowledgements

I would like to express my great appreciation to all the people I have come across during my career. Senior PM´s, co-workers, architects, various experts within their special field, maintenance workers, representatives of city councils plus friends and family. Especially my family and friends who have come up with good points during the production of this book.

Author's background

I have been in the construction industry for the past 20 years, for 8 years I worked as a carpenter but changed my profession to project manager in my late twenties. My project management career is now in its 12th year. Initially I started off with small projects, such as project managing office rebuilds, but was quickly given more responsibility in building condominiums, train and bus depots. These days I work as Business Unit manager at a larger consultancy firm providing expertise within infrastructure, industry, buildings, energy, communication and security.

I hold a Bachelor's degree in construction management, MBA degree from the University of Liverpool, PMP, RMP and ACP certificates from the global credential Project Management Institute (PMI).

Introduction

If you are seriously contemplating to create your own home, this is the book for you. Its purpose is to guide and help you through the process of one of the most daunting tasks you have ever ventured into. Acting as a construction manager of a Self-Build is not just any project, it is your new home in which you might live for the rest of your life.

This book will help and guide you through the time consuming and complex process by raising your awareness of all the pitfalls and challenges that might arise, providing you with the necessary tools to successfully act as a construction project manager. It will cover important issues such as design, contractors, money saving tips and keeping relationships intact. Doing things right from the start will indirectly remove a great be deal of stress and a successful finish of the project is more likely to be achieved. At best it will be an exciting journey resulting in a great sense of pride and satisfaction.

Tips presented in this book derive from personal experience when acting as a construction and project manager combined with the tools taught by the PMI (Project Manager Institute). This book will also highlight the need of understanding stakeholder management (people with various interests in your project). In this case it is about the emotions of the people involved and how to deal with assumptions and expectations. Not all family members have a stress level that can deal with a high level of uncertainty and setbacks, so to make the project successful it is crucial to make sure everybody involved is aware of and mentally prepared for the demanding task to build a house.

Project management can be very advanced, but also quite simple. My strategy has been to produce a book at a level that will be easily comprehensible to anyone who is not accustomed to the role of a project manager. It will cover a project that is "normal", in the sense of size, budget and a type of construction that most people go for.

This book can also be for those of you who wish to get a better understanding of what kind of challenges there are ahead, when it comes to project managing a housing project. A project manager's way of thinking entails unfolding the unknown to ensure a realistic and correct planning.

This book will not address all aspects of project management in detail (advanced project management), it is more about helping you to adopt a realistic mindset of construction project management with the help of some necessary project management tools. It will also advise you on how to deal with a lot of activities during the project. If you have grand plans e g to create an exceptional garden, I recommend you explore books on "how to create a garden".

The "easy" way of realizing the project is to get a fixed price from a contractor who will complete the whole house. By doing so, be assured there will be a great deal of contingency reserve set aside by the contractor (depending on the uncertainties). Sometimes as much as 20-25% of the entire budget. This book is not about that, this book is about helping you and your partner(s) fulfilling the task of managing your project all the way in a successful manner and keeping your relationships intact until the end. This book is undoubtedly for you thinking that planning is a waste of time.

The currency Euro will be used in the book. The assumption that 1 Euro = 1 US dollar = 1 Pound. The currency and costs presented in this book should in this context merely be viewed as a presentation tool for discussion purposes.

An Idea/Vision

Many people harbour a vision of a dream home tailored to their needs, tastes and life style and you are probably one of those who have entertained the idea of finding a plot where to custom build to satisfy your and your partner's or family's requirements for a future home. Once you and your family share a vision what the house should look like and agree on realizing this vision, then the first little step is achieved which is great.

Now the moment has come to start exploring the possibility of realising your vision. Before you literally start realising the vision, as in hands on, there is a lot of work to be done. Above all, a great deal of planning must be done to see if the vision you have in mind is realistic to start with when it comes to the desired quality, time schedule and budget. Another aspect that should be considered at this stage is how to keep you and your family intact while you are trying to complete the project. Building your dream home should be filled with excitement and joy while you are in control, not causing stress.

Inspiration

You and your partner probably have a good idea about the house design you prefer before doing any kind of feasible analysis. At this point, however, you have not considered all the details that are required in order to succeed with the project. An inexpensive way to get inspiration is to visit hardware stores, buy a few magazines, look at certain TV programmes about homes and architecture, sign up on Instagram, Facebook and let yourself get inspired by the latest solutions, products, architectural features or interior designs on the market. Use a notebook or equivalent and note what specific characteristics the specific product has. Was it expensive, easily installed, what did it need to function, etc? Ask yourself questions at this point that the manufacturer might omit. Manufacturers obviously point out the positive aspects about their products but pay less attention to what really matters production wise.

Planning

The importance of planning cannot be underestimated as it has such a huge impact on how we perceive assumptions & expectations. Planning is all about minimizing uncertainty from the project and thereby enabling you to draw up a realistic budget and time schedule. Equally important it is to bring alignment to you and your family's assumptions & expectations when it comes to quality. Listing and agreeing on your and your family's requirements can take time. It consists of choosing type of house, floors, paint colour, various installations and fittings, exterior walls, windows etc. Parallel with compiling requirements, the need of understanding the cost becomes apparent. The greater reliability, thanks to planning, the less stress for you and your family. At the same time contingency expenditure (unknown risks/surprises) will be kept at a minimum for the project. Furthermore, with planning you will avoid making any last modifications during production (changes in the project in order to save money) and thereby avoid a solution you are not really satisfied with. It is also during the planning stage you will be organizing the activities and analysing what risks there are in your project and how the risks will be dealt with.

Planning takes time, but the time spent here will allow for a smooth production, and it is during the production stage most of your funds will be spent. That is why planning is such a key factor. Otherwise you might have to make quick decisions at a given moment during production which will not be thought through resulting in something you and your partner not really want.

Finally, it is important the whole family are involved in this process and fully understand and agree on the decisions made during planning! Below is a visualization of planning, the more you plan the more exact everything will be before production start! But do not worry, you will be presented the tools to use later on in this book!

Planning

Quality – The house, what do you want and what is feasible within your budget time schedule?

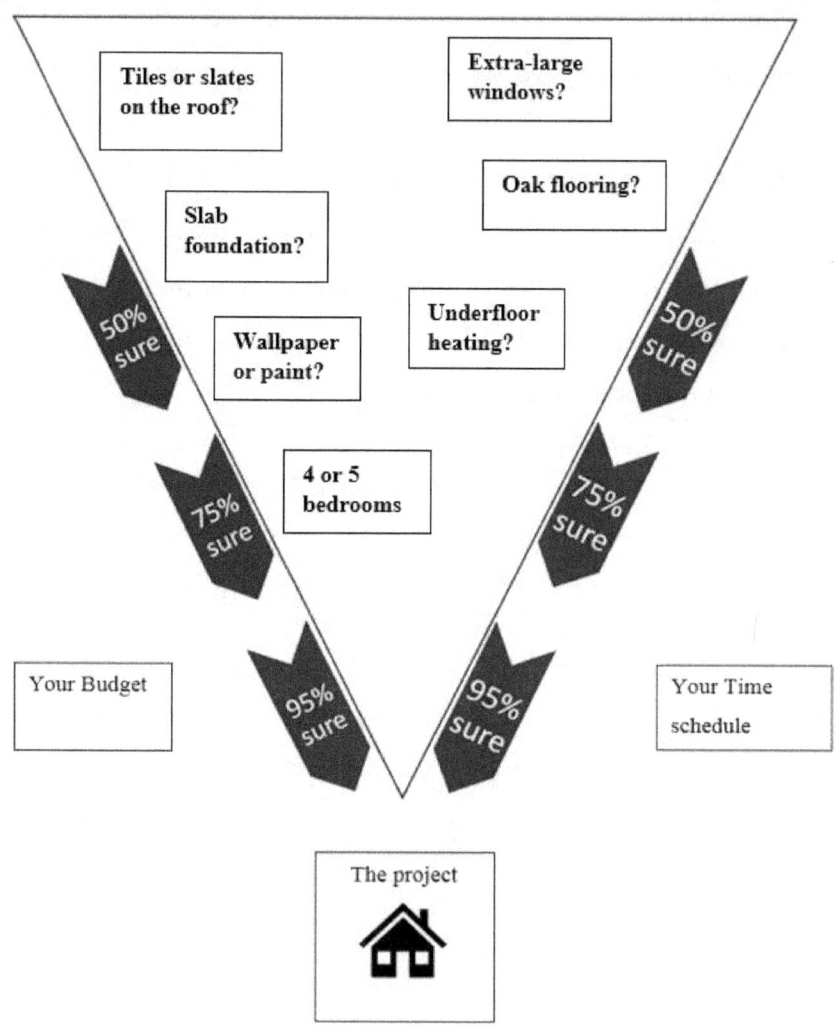

Assumptions & Expectations

A few words about assumptions and expectations, assumption is a thing that is assumed to be true and expectation a belief that something will happen or be the case.

Discuss assumptions and expectations with your partner so he or she fully understands what the project entails, including costs and everyday activities in the project. Assumptions and expectations should also be discussed with the contractors to ensure they are aligned with yours. This should, however, be done during planning so the agreement/contract will be the future reality fulfilling what you both envisioned. These talks should be ongoing throughout the project.

Perhaps you have a good understanding of what is illustrated on a drawing and you can easily visualize what the result will be like. But do not take for granted that your partner can do the same. It might be worth having the architect or an interior designer make a few drafts in 3D for both parties to comprehend what the drawing illustrates. By doing so, assumptions and expectations will be aligned, hence mistrust and stress will be reduced.

Discuss the most important issues together, write them down and explore the best way to bring clarity. Ask questions to really find out what assumptions and expectations your partner has.

Interpersonal skills and the use of a structured working approach, being thorough and asking questions are key factors to achieve good communication. Not only does this apply to your family, but also to the contractor(s) and other people who will be involved in your project.

Apart from discussing the end goal, it is necessary to discuss assumptions and expectations you might have regarding the production of the project and what everyday life will be like during the project. A good way to get your assumptions and expectations aligned with a contractor is to visit reference projects, perhaps a previous build by the contractor or a friend's house. Both parties will then know exactly what to expect!

You might for instance think it will be easy and cheap to change something in mid production. It is not. You should aim to be 95-98% clear and have everyone aligned as to how, what and when activities will get done once you leave the planning phase.

Here is an example when expectations between two parties do not correlate with each other:

Let's say you ordered large windows for your project and perhaps you think it is obvious they will also be fitted by the manufacturer once they are delivered – while the manufacturer on the other hand only left a quote for delivery. Or perhaps you booked a crane and labour to fit the large windows once they arrive, only to realize the length of the crane's telescopic boom is not adequate and therefore does not fulfil its purpose. The consequence of a situation like this, will most likely result in you having to pay for labour and crane since you cannot store the windows due to the lack of space at the project site. Having to send the windows back to the manufacturer will mean unexpected expenses since the manufacturer might charge you for storage, too. A potential loss of several thousand Euros in one day will have a bad impact on your budget and time schedule! Avoid this situation by planning and by asking the supplier you deal with the right questions. Communication is such a crucial aspect and one of several keys to success!

Notes – What assumptions or expectations are not the same with your partner or contractor?

Feasibility Analysis

Before you **start** the project, a feasibility analysis is required of the vision you have in mind. Some basic questions and issues should be answered and considered:

- Cost for the plot?
- Is it likely to get a planning permission from the city council for your chosen house you want to build at the plot you are looking at?
- Is it possible to install electricity, water and drainage? What is the cost?
- Fibre or copper for the internet? What is the cost?
- Is there a company who will collect rubbish at the plot? (Maintenance cost)
- Will a rough time estimate of 15-20 months before moving in be realistic?
- Are there good contractors in the area who can help you?
- Estimate the cost and draw up a rough budget and decide whether to go forward or not. You could for instance use a cost per square meter in order to decide if it is worth proceeding. A rough estimate of the building cost per square meter is perhaps 2,500 Euros in the area where you want to build. If your vision of the project is 200 square meters, a rough amount of 500,000 Euros + plot would be the project cost. Is the sum feasible for your family's budget? Engage a few contractors who you might hire and talk to them about the square meter cost and time frame.

- Look at other recent building projects in the area and get references at the same time.
- Consider your choice of location as it could have great impact on your future calculations. If you are new to the area it would be wise to speak to the locals, local journalists and the council about future development plans that might affect your project.
- Consider your temporary accommodation during the build and the possibility to commute.
- Is the project site easily accessible? Check the width and quality of the road. Are there any tunnels or bridges that might cause problems for transports?
- Do you have a need or wish to cycle to work, is that bearable/possible?
- Are the distances to work, shopping facilities and day care acceptable? What is the traffic
- like in the morning and in the evening? Will it be manageable for you and your family?

Make a checklist if needed, to ensure you have not missed any vital information at this early stage before proceeding. A good idea is to consider help from an architect early on to make a draft of what your desired house design. When processing the architectural vision, you must know your needs in terms of space needed and what budget you dispose of. Keep the initial draft of the drawing rough and flexible so you can more easily adjust it to the plot, if possible. With the help of a contractor or perhaps the architect, you can get an idea if it is feasible to begin with by knowing the square meter price of producing a new build. It will give you an estimate of what the envisioned project will cost. If you and the rest of the family are happy with the cost and other factors such as distance, public transport etc, then go ahead. Move into the new phase, the phase of detailed planning.

A suggestion is to give you and your family one year of planning, but then again, it is all about how advanced your project is, how much time you have and so on. Most importantly, do not rush when planning!

Planning Phase – Plan, Plan and Plan

When entering the planning phase, an assumption is made that you have acquired a plot and have an answer to all the questions you could think of in the feasibility analysis and you and your family are all in agreement with each other.

During the planning phase you and your family are going to decide on the many requirements you have listed, such as the design of your home, the size of all rooms, the height of the ceilings, what flooring goes into each room of the house, the material to be used, installations, fittings etc, etc. Most importantly work through the calculations so you can create a realistic and sustainable budget and time schedule for your project. There is a lot of information that needs to be gathered and considered. Involve contractors and sub-contractors early and discuss the planning level required, what they seek. But it as to correlate with what you want of course. The more thorough job you do in the planning phase, the better and smoothly everything will proceed during production, hence your budget and time schedule will be intact. So, do take it seriously and do your absolute best!

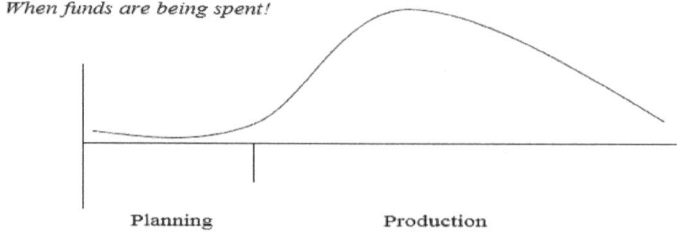

Notes – Write down what "things" you think need extra planning:

Break the Project Down

To get a good overview of the things necessary to create your dream home you should "break it down" into categories and activities to get a comprehensible overview of all the stages and activities necessary to complete the project. You do not, however, need to break it down to the smallest detail.

Decisions on what kind of wiring, concrete, rebar or plasterboards that will be needed, will be made by the specific, hired contractor. If a particular detail is of great importance to you, then of course you need to ensure it is mentioned in the contract. While looking at all the activities, you should simultaneously consider what the specific activity means in terms of cost, time and quality standard.

Furthermore, you will also have to decide on the positions of the sockets, how many, switches and what products and colour you want, which way a door should open, where in the ceiling a lamp should be placed. Additionally, you will have to analyse what activities intervene with each other, such as electrical wiring, ventilation and plumbing that you might want to have inside your walls. With the help of drawings, discuss this with each trade (electrical, plumbing, ventilation etc) to ensure that they have a plan and can work without any interruptions.

By breaking down the project in categories it will ultimately enable you to draw up clear and lucid agreements with various contractors. The contractors will also be able to make a more exact bid, as the extent of the work is clear due to the planning you have done. Thanks to splitting the project into smaller parts and providing answers to the key questions who, when and how, the bid(s) will be more realistic.

WBS – Work Breakdown Structure

WBS – Work Breakdown Structure is the foundation of the project no matter how big or small, advanced or simple it is. It will help bring structure and define scope and should be used when you plan for your project, no exceptions! It will give you a great overview of all the activities that needs to be dealt with and of great help while creating the budget and time schedule for your project.

WBS also:
- Helps to identify risks
- Helps to prevent activities being missed during production
- Provides evidence of the need for resources, funds and time
- Facilitates communication and cooperation between the people involved

Activity = Work Package

WBS is structured in so called activities, also known as **work packages,** and should be applied until all activities are achieved. If the activity estimate is realistic it should not exceed more than 2-20 hours of work. Involve your friends and family when structuring your WBS to avoid missing any activity. An example of the activity "painting the living room" should take no more than 20h. Below is an example of a WBS done in Microsoft excel format. You will get access to the excel document through the www address on page 86.

The more details you provide in the WBS, the clearer the project will be to everyone involved!

ROOMS									
EXTERIOR WALLS	ROOF	WINDOWS	INNER WALLS	Living room	Bedroom 1	Bedroom 2	Bedroom 3	Kitchen	
Exterior wall 1			PAINTING/WALLPAPER	Living room	Bedroom 1	Bedroom 2	Bedroom 3	Kitchen	
Exterior wall 2		Living room	INTERNET	Living room	Bedroom 1	Bedroom 2	Bedroom 3	Work room	
Exterior wall 3		Bedroom 1	VENTILATION	Living room	Bedroom 1	Bedroom 2	Bedroom 3	Kitchen	
Exterior wall 4		Bedroom 2	PLUMBING	Kitchen	Bathroom 1	Bathroom 2	Utility room	Exterior wall 1	
		Bedroom 3	ELECTRICITY	Living room	Bedroom 1	Bedroom 2	Bedroom 3	Kitchen	
						Corridor	Work room	Hallway	

Clarifications of several WBS examples given:

- Activity roof is under Activity windows because there will be a fitted window on the roof.
- If the garage is detached from the house, then it should be treated as a sub-project to the actual house, therefore a sub-WBS should be created with more or less all the activities you have under the housing project.
- Exterior walls are marked 1,2,3 and 4 because of all the sides of the house and each wall could have different materials such as bricks or panels therefore the example given requires further division.
- Under Plumbing you have exterior wall 1, where you might consider a hose connection for future garden use or perhaps when cleaning the car.
- The WBS will act as a basis for the contract documentation with the contractor eventually. Under the chapter Agreement, the WBS will be discussed further and its role when the agreement is drawn up.
- Doors, floor moulding trim and tile edging trim are left out in the WBS. Use the SOW (Statement of work) for that kind of information, if required.
- The garden might need electricity for exterior lighting and outdoor decking. Further divisions are required.
- If you are unsure which rooms need ventilation, discuss the matter with a ventilation contractor.

An example of not being too specific with requirements:

If the foundation of the house is a concrete slab for instance, there is no need to break it down further than stating the required width, length and depth. That information will be enough for a contractor to calculate the material and labour needed. If you wish plumbing, electrical wiring or underfloor heating in the slab, then that is information necessary to be passed on to the contractor so he or she can make a precise bid. Something that is left out, due to the fact it should be so obvious, is the quality of the concrete and the amount of reinforcement. Not mentioned here, but later on, is the possible need of piles and other groundwork before building the slab. Acting as a project manager you will have to ensure that the sub-contractors (electrician and plumber) coordinate and finish their work by placing the ducting, drain and underfloor heating before putting the concrete in. Communication is key!

Here is an example of factors and materials when laying a slab (something you do not need to get your head around):

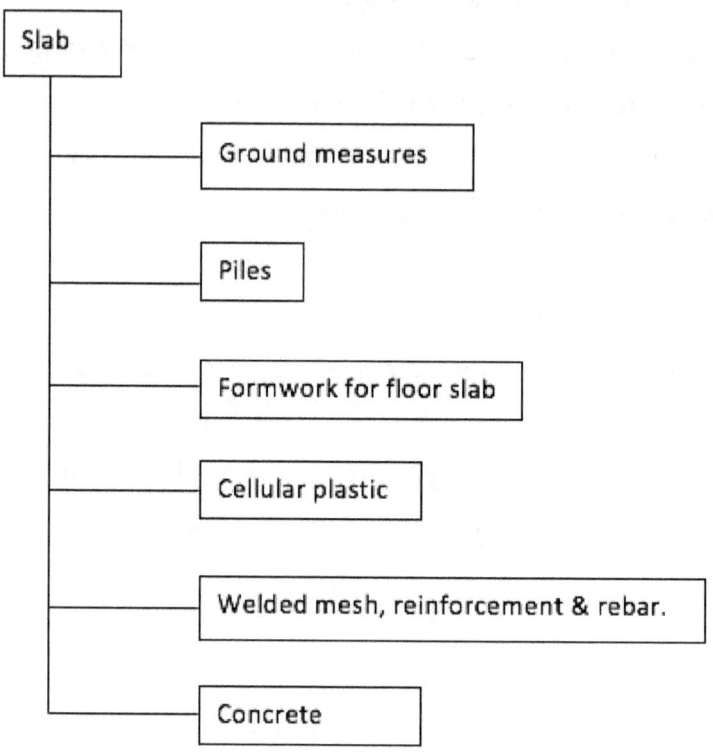

The uncertainty left is ground measures therefore it is important to bring in a geotechnical expert to provide a report, based on tests of the ground where the building will be situated. It will provide answers (not all answers) regarding rock level, boulders and how to build a solid foundation for the house. All this information will be needed when producing the procurement document SOW (Statement of Work). This will be discussed later in this book under chapter Agreement. If you decide to build something out of the ordinary, then you might have to speak to a structural engineer.

There are several documents that will aid you to get a better understanding of all the things that need to be done, documents that will answer What, How and When. Once you have a vision of what you want to create, you need to produce these documents with the help of an architect, as the architect knows what space you need in order to make your future home habitable and at the same time discuss your needs. The documents will be discussed under chapter Agreement.

The Organization – The Team

Who will do what? By creating a simple table of those involved in your project, you can easily correlate the specific contractor with the activity in the WBS and timetable.

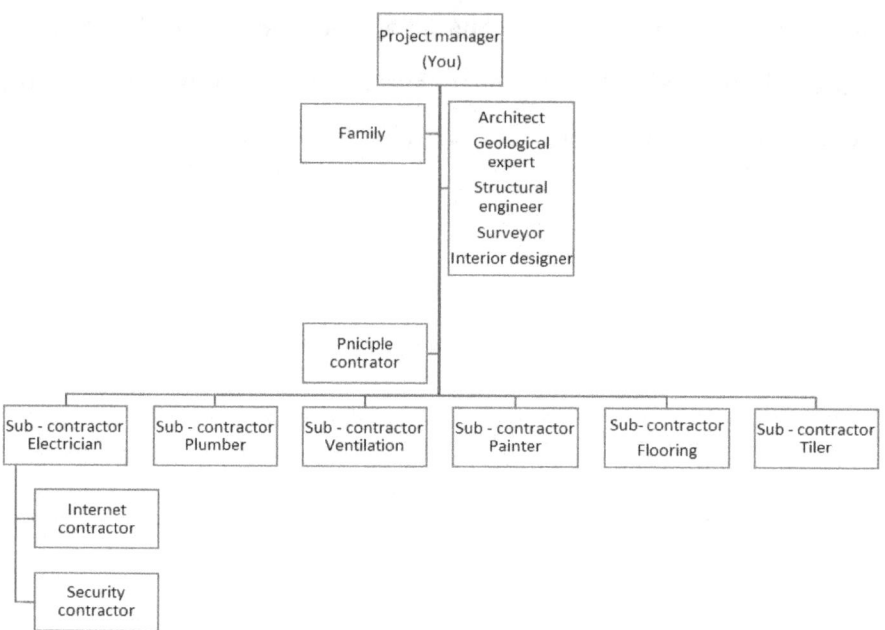

The idea presented here is that the principal contractor does the slab, exterior walls, inner walls and roof. You and the principal contractor collaborate closely, but you have the main responsibility to coordinate the principal contractor and sub-contractors. This is, however, a task you can delegate to the principal contractor in the agreement if you feel overwhelmed. I have also put a couple of activities under sub-electrician so he or she can be responsible for those installations.

As an initiator of a project, you will also have to understand what responsibilities you have regarding health and safety or any other building regulations. Ensure you are aware of all obligations by talking to various people. Preferably you want each contractor to be responsible for the health and safety for their workers.

Notes – which contractors are you going to use?

Choice of Production Method / Finish

There are several production methods for a building, some of them are more expensive than others. What kind of production method and materials you choose will have a significant impact on your budget. Bricks for instance, are more expensive than render, glass is more expensive than an exterior brick wall, asphalted cardboard roofing is cheaper than sheet metal on the roof. At an early stage of planning, you can quite easily decide what approach to use and what material to discard or change. It does however have to correlate with the architectural vision you have in mind.

For instance, one decision that has a significant impact on design and cost, is whether you decide to have visible electrical wiring and plumbing instead of having the installation go inside the wall. Choosing this kind of design and method will make it easier for the production to progress, since every activity has been isolated from the other, as seen below:

1. Walls done
2. Painting done
3. Wiring done
4. Plumbing done

There is no need for coordination between the various contractors and work will proceed more smoothly. The downside is the obvious visible wiring and plumbing. Aesthetically this method is questionable.

Apart from the method of production and the design of your envisioned house, it is equally important to think about the inside of the house, not only the finish but how the house will fulfil your needs, not just from a practical side, but also from an emotional point of view.

By making a rough plan of where to put your furniture and ornaments, questions are likely to arise. Perhaps you want a large painting hung or a grand piano in the living room. Are the windows or doors large enough to accommodate this request of yours, will the items fit so to speak? The earlier you start thinking about "all" the items and the emotional aspects of the house the better. These are essential issues to discuss with your partner.

Prefabricated Products

Depending on what your intentions are with the project, and what you are trying to build, prefabricated products might be right for you. Imagine having exterior walls delivered with already mounted windows, trusses and sections of tongued and grooved boards to the roof. These products would enable you to make the house waterproof within a few weeks, allowing you to continue working on the house in a much better environment, i e dry. Even kitchen and bathroom furniture can be prefabricated. Production methods like these are widely used today and definitely worth looking into. The initial cost might be higher with prefabricated products, than doing everything on site, piece by piece. The benefits, however, are shorter production time and thereby lower temporary living costs. Other benefits that come with a shorter production time are all the other temporary things your family will have to endure during the project.

Think Things Through – Avoid Implications

A drawing showing the positions of spotlights in the ceiling

Your desired bathroom design

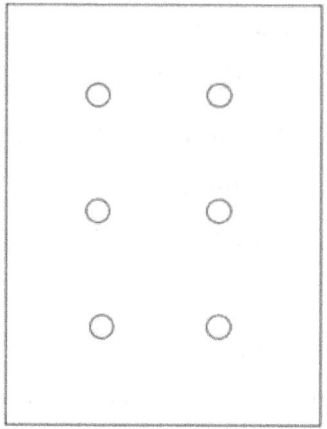 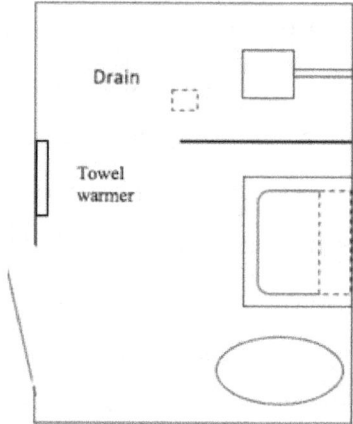

These drawings look quite good at a first glimpse. The spotlights look all symmetrical in the ceiling and the placement of the bathroom equipment looks good. But if you were to put the ceiling drawing on top of the bathroom, you will see that the spotlight top right will be "hidden" by the shower faucet. It will likely create a huge shadow in the bathroom unless it is moved. This is just a small example of collision between two trades (plumbing and electrical). Not only that, if you would like to get good light on your face, you would probably want a mirror with built in lighting, so the light can get directly on your face which means you have to choose the mirror early in the planning phase and ensure that the tiler and electrician speak with one another.

What you do not see here is under floor heating and if you would like an advanced bathroom design with e g a symmetrical tile pattern, it would be wise to ask an architect to produce such a wall drawing to avoid mistakes and clarify where everything should go, such as furniture, shower faucet, switch, etc. You can for instance just choose pictures from a magazine and use them as reference to achieve the design you want in order to save money.

There are many things to consider, not just from a collision perspective but also from a needs perspective.

Bedroom for instance:

What? Where? Do you need a place for a phone charger? Lamp when reading? What height? Led lamp in closet? What manufacturer? Airconditioning? What kind of window? Windowsill? TV on wall? Internet? Type of closet? Ask yourself the right questions and write the answers down.

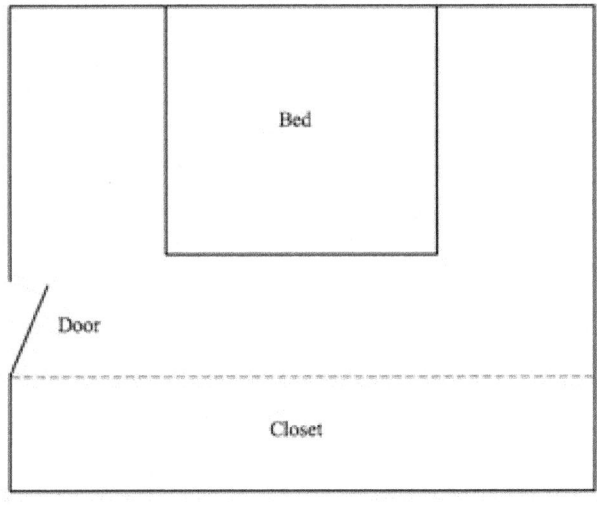

You will also have to decide what kind of flooring you want/need, what finish you like on the walls and the ceiling. With the help of the document "room description", discussed in chapter Agreement, you write down exactly what you want together with SOW. If you find it complicated, you can always consult with interior designers who are experts in this field.

When it comes to the kitchen, the makers of kitchens are usually great at giving tips and pointing out things you might want. It is their job to meet the customer's needs.

Notes – What specific needs do you have? Must have, want to have, could have:

When to Start Production

When to start production needs some consideration. Obviously, you need to talk to the principle contractor to find out when he/she will have time and labour to start and what weather conditions are most favourable to start laying the foundation. If you live in an area with a rainy season or winter season, the groundwork and foundation may cost 10-30% more, due to the difficulties the contractor may face because of bad weather conditions. So, make sure you plan production at an ideal time to ensure that as many factors as possible are in your favour. The same goes for applying render. Some companies require at least 5 °C to work with their product to be able to give a guarantee.

Apart from ideal production conditions consider what would be ideal for your family once the build is initiated. Obligations such as temporary accommodation, transport to school, etc must be met.

Visual Details

With the document room description, you clarify to yourself and others what the end goal will be in each room. To meet your expectations or bring clarity to other questions we are going to look at details:

- Colour, material and manufacturer of outlets and power points?
- Colour, material and manufacturer of toilets, sinks, faucets, dryers, washing machines?

- Light bulbs, regular or led, colour of the actual light?
- Wall hung bathroom furniture, toilet?
- Hidden plumbing?
- Visual or hidden radiator plumbing?

Once these questions are addressed, write them down in the room description (product and identification number) or in the SOW. While planning this, you must constantly consider what the details entail in terms of cost and duration once in production.

Understanding Correlation between Activities

When you make your decisions on what building materials to use there will always be pros and cons with the specific material and production method. I recommend that you write a list of pros and cons for your own sake. While doing that, involve contractors or manufacturers or even sellers at hardware stores and discuss your list in order to get more information about the specific material or production method you are considering.

Activity: Underfloor heating in slab

Pros	Cons
Energy efficient heating	High initial cost
No radiators are needed (unless you have more than one floor)	
Safe and comfortable heating	

Activity: Triple glazed window

Pros	Cons
High insulation capacity keeps loss of energy at a minimum.	High initial cost

Activity: Exterior wooden wall panels

Pros	Cons
Cheap compared to bricks	Maintenance, needs re-painting every 10 years

Making a list of pros and cons will raise your awareness and ability to avoid surprises and give you more confidence when going ahead with your decisions. Apart from a pros and cons list, it is important to ask yourself: What does the chosen production method or installation require in order to function 100%, once the work is completed? When fitting the kitchen for example, there will be several trades involved to make the kitchen fully functional:

1. Ensure that all necessary installations are in order such as water, drainage, electricity and ventilation.
* Painting
* Flooring
2. You will need a carpenter to assemble and fit the kitchen and perhaps a tiler
3. plumber for fitting the tap and dishwasher
4. electrician for installing the oven, dishwasher, refrigerator, freezer, lights
5. Fan fitter

A common pitfall on a building site is the idea that having several groups of craftsmen working in the same area at the same time will hasten work progress. But that is rarely successful, since their work will be very inefficient. Cause every labour needs space for himself, co-workers, material and equipment. Therefore, your planning effort is important together with coordination and communication.

People Involved

There will be a lot of people involved in your project like members of the local council, the principal contractor, all the sub-contractors such as electricians, plumbers, foundation contractor and of course your family. Everyone will require different information from you so they can help you in the best possible way, therefore it is important that you involve them and ask what kind of information they seek.

It is a great advantage to have an extrovert personality which facilitates dealing with all different personalities and to be assertive when required. There will be situations when agreements can be difficult to reach. All too often people have too much on their plate and forget, or just do not prioritize you. But by being pro-active and calling 1-2 times extra to really ensure your delivery will be there on time or for example make sure the labour force will show up at a particular point will be worth the extra effort. It is all about what kind of collaboration you have been able to establish with the hired contractors.

Family

You are likely to involve your whole family in one way or another and even friends and acquaintances in your project. This is something that is rarely addressed to the extent needed. Despite the project in progress, life will continue as usual. Leaving kids at the day care or helping them with their homework, shopping at the supermarket and walking the dog are chores that still will have to be carried out. It is also of great importance to have an understanding for the emotions that are affected by the building project and to be able to show your loved ones affection and support in the process.

This can be a scary feeling for some! After all, we are creatures of emotions. By acknowledging that fact, one is likely to be more tolerant and understand why people involved might behave in the way they do. Do not forget that communication plays an important role to address this!

Discuss this with your partner and find a consensus for the approach forward, how you will deal with all the activities going on beside the actual project.

Contractor and Sub-contractor

Depending on what kind of agreement you have with the main contractor and sub-contractors, dialogue and collaboration are always important, since both parties should "buy-in" to the project and help it succeed. A "Buy-in" is when a person or company actively agreed on how things should get done. Most contractors or sub-contractors have 4-5 projects going at the same time, because the craftsmen cannot be at the project all at once. And the employer will have to make sure the employees have jobs to do all the time.

An example to illustrate this: before the electrician does the wiring in the wall, he or she is dependent on the carpenter building the wall frame and perhaps plastering the boards on one side of the wall before doing the actual wiring. Therefore, it is always important to have a discussion with everyone involved to fully understand why the sub-contractors need to collaborate with each other. By doing this it will be easier to understand and create a more realistic time schedule and get a better understanding of the work required.

Having so many people involved, it is likely that some will fall ill or be affected by unforeseen events that will make the planning slip a bit. This is something you just cannot do anything about, just accept and embrace the fact that things happen that neither you nor anyone else has any control of. This is something you will look more into in the chapter Risks.

You will also be likely to experience the fact that the contractor will ask for more time during production, like the finishing of the facade for example. Consequently, the rent for scaffolding will increase.

One way of dealing with this situation, for your own sake, is to get a fixed price from a contractor for a specific job. By doing so, you have no risk as you most certainly have transferred the risk to the contractor. But, if one contractor takes more time, in might inflict another contractor who will do work afterwards. That is why it is important with a bit of space (time) between contractors' work. Once you have decided which contractors you are going to work with (carpenter, electrician, plumber, ventilation fitter) share and discuss your ideas and thoughts concerning the project with them. Be open minded, perhaps the contractor and sub-contractors have a more effective solution than you have regarding your end goal.

These kinds of work discussions should be had before production even started, not after, since changes in mid-production can be costly.

Never pay contractors upfront, just pay for work carried out, because if the contractor goes bankrupt (for unknown reasons), you want to be in a position where a new contractor can take the former contractor's place without losing too much momentum and money in the project.

References – the Contractor

When hiring a contractor, do not just hire anyone. Look beyond the price tag! Make sure the contractor has done similar work previously to yours. Visit the contractor's previous projects and speak to the people who hired him or her. Remember, the contractor is someone you will be working with very closely and who ultimately will be a key player when it comes to delivering the project and meeting your expectations in terms of quality, expected time and budget.

Preferably you want a contractor who is very understanding, having a high EQ (Emotional intelligence) level and understands your needs and will hopefully be your "partner in crime" while progressing through the project. Making the effort to know the contractor you are going to hire will pay off in the end. If you feel safe with the contractor and confident of good collaboration the whole family will benefit.

When you gather information about the contractor (from references), ask the following questions:

1. What kind of work did the contractor do for you?
- Were they acting as the principal contractor building a new house?

2. Was it easy to align your vision with the work that should be done by the contractor?
   - Were there many discussions about changes etc from the contractor's part?

3. Did the workers show up on time?
   - Did they turn up as discussed?
   - Where there a lot of excuses?
   - Did they tidy up after themselves?
   - Did they always do the full hours?

4. Look up the contractor's credentials.
   - Is the contractor required to have a license?
   - What is the contractor's financial situation?
   - Is the contractor insured?

5. What was it like to cooperate with the contractor or the subcontractors?
   - Was it easy to have changes made?
   - Did the contractor take the initiative to suggest ideas that benefitted the project?

6. Was communication with the contractor adequate?
   - Did you have frequent meetings with the contractor to discuss time schedule, budget, changes etc?

7. Did the contractor complete the work on the agreed budget?
- If not, what were the primary reasons for this?

8. Were you happy with the end goal? Did it meet your expectations?
- Did the contractor fulfil the obligations according to the agreement?
- Did the contractor complete the work on time?
- Did the contractor pass the preliminary and final inspection and fix any problems straight after?

9. Would you hire this contractor again?
- If not, ask for a detailed explanation. What exactly was not to your satisfaction?

Other actions to take:
- Always get at least three bids.
- **Never agree to do upfront payments. Never!**

References – the Consultants

<u>The architect</u>

When you hire the architect, you want to see the previous work done and talk to previous clients. You want to get an idea if the clients were satisfied with the work done, what the work costed and so on.

<u>The Geological engineer</u>

Here you want to know the cost of course but more importantly, was the produced documentation about the ground satisfying for the contractor to work with. Get references from previous projects and talk to the involved contractors. It is crucial that you get this right, since the work regarding the ground can have a huge impact on cost.

Environmental aspects

There are several things you can add to your project that will be beneficial in the long and short run from an environmental perspective. However, installation of e g solar panels or installation of an electrical charger for your car will most likely come at an initially high cost. Or if you have a desire to reduce future heating costs, tight building and proper insulation of exterior walls will be costly but reduce the heating cost in the long run.

Some authorities or financial institutes, depending on where you live, may subsidize loans or give discounts to your project if you get an environmental certificate. That is worth looking into, not just for your finances but for the environment, too, of course.

Temporary Living

Your temporary solution for accommodation will depend on your budget and should allow you and your family to have a reasonably "normal" life during the project. Therefore, your temporary solution plays an important role for the level of stress within the family during the build, so think this through!

Make sure the time spent at the temporary accommodation is overlapping the date when the project should be finished to avoid a raised stress level. Do not plan to move out from your temporary living on the exact date your project is planned to be finished. Do not plan to move into to your new home until it is 100% completed. You should also avoid having to find yourself staying in a crowded place together with your in-laws, or in too small a caravan on your project. Temporary accommodation will have to allow life outside the building project to go on as usual.

If you plan to sell your present home, you can always try to sell it when the market is most favourable. Discuss this with a few local real estate agents, what time of year is the best to sell.

Drawings

The following drawings should be produced by your architect:

- Floor plan including windows, walls, floor
- One or two sections drawings which will show the height
- Facade drawings of every side of the house
- Roof

These drawings are the bare minimum. With the floor plan you can create a separate basis for other trades such as plumbing, electricity, ventilation, interior. Always use the same scale on the blueprint to ensure realistic measurements, which is something the architect will be able to help you with.

See example below:

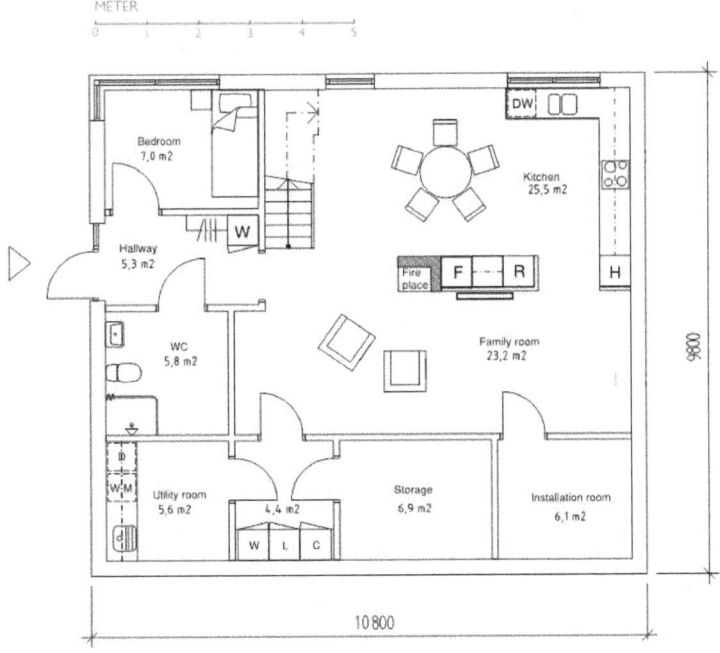

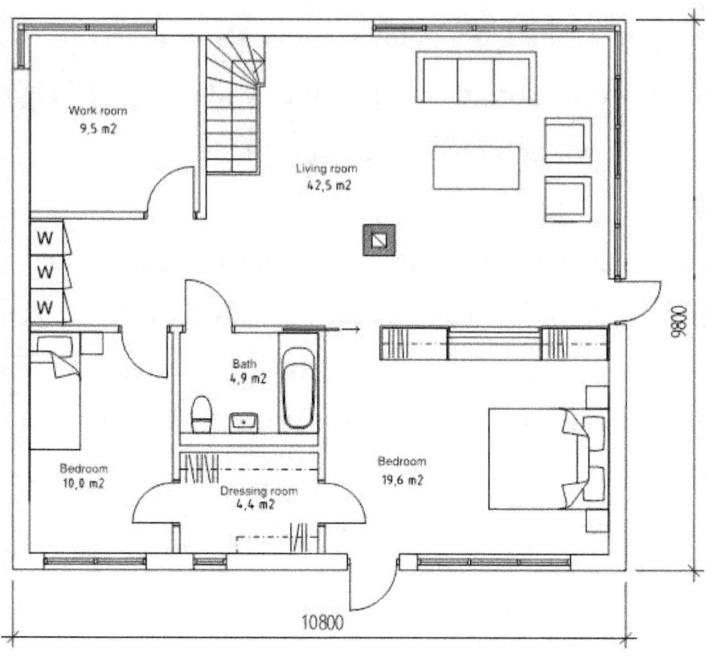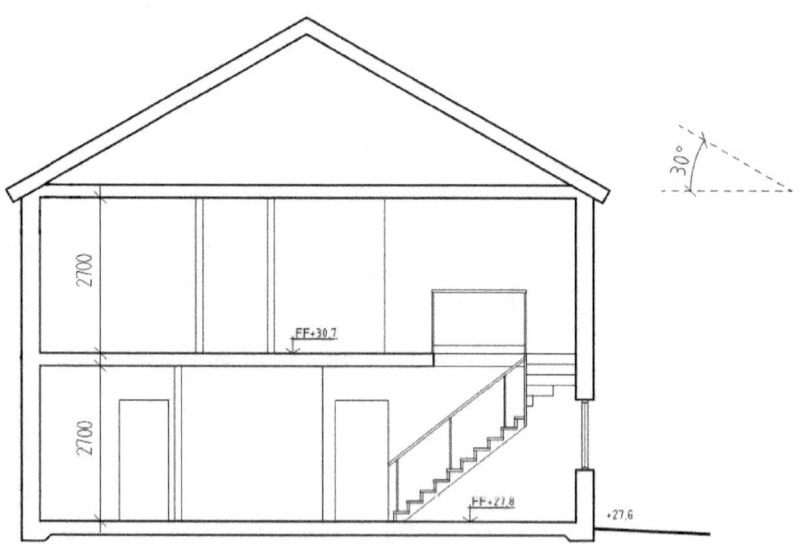

Notes – Have you asked the architect to draw all necessary details to the contractor so he or she can see what needs to get built?

Here is an example when producing information towards an electrician:

Let's say you want to be specific to one of the sub-contractors who is responsible for all the electricity. With the floor plan, you can point out where you want the sockets, switches, internet connections, spotlights and other parts that the contractor will install.

Let's see a part of a simple drawing of a floor plan intended for the electrical contractor:

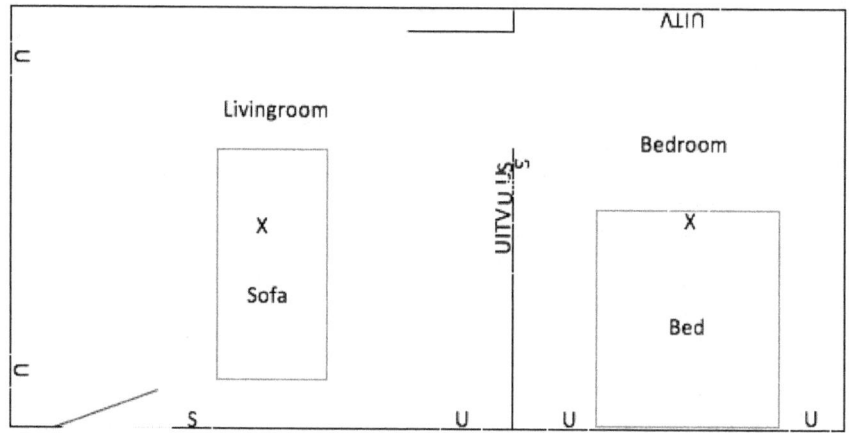

X = Socket for lamp in ceiling	I = Internet
U = Socket	T = TV cable
S = Switch	

With this information the electrician can easily calculate the amount of work needed to get the job done. In addition to the drawing, you might want to give a general height for sockets and switches and what brand you want them to use. This information you can describe in the SOW. Also make it clear that you want all wiring to be hidden inside the wall.

The architect should produce drawings of each wall, floor and the ceiling if spotlights will be used. The drawings will be used as the basis in the agreement with the contractor, so there is no confusion as to what is expected.

Depending on what kitchen you want, it is most likely that the manufacturer, who will deliver your kitchen, will supply all the drawings. Your job will be to provide the correct measurements and position where the future kitchen will be.

Early Planning Done

At this point you should have done the following:

- A preliminary WBS, an idea of what material(s) should be used in the house. A rough idea as to who will do what and when, and an initial structure of the organization planned!
- Created a budget in relation to the WBS.
- Created a time schedule in relation to the WBS and have a good idea when to start production.
- Talked to a few contractors and sub-contractors about your project and found out how they work and what information they want prior to helping you out. More importantly get a feel, how they behave and perhaps even look at some reference work they have done.
- Have an architect involved who has made drawings of the whole house.
- Have a rough idea about the architectural features.
- Decided where you are going to live during the build.
- A list of what installations you want, but not necessarily boiled it down to, "must have", "nice to have" or "convenient".

Planning will now move on, exploring all the opportunities that might benefit your project, your family needs and budget and as well as time schedule.

The Project/House itself

When it comes to the house itself, there are obviously many things that need to be covered so the end goal meets your vision and needs as fully as possible. You are likely to go back and forward in the planning phase, compromising with your partner what the end result will look like, and at the same time ensuring that cost and time do not disrupt your set budget. You realize e g that the kitchen you want is much more expensive than you thought, but there is a manufacturer who can produce a similar kitchen at 25% less, which will give you the opportunity to purchase expensive appliances (that you really want) and so on. It will be a give and take situation which will be ongoing, when discussing the end goal with your family. It is all about finding a sweet spot of your family requirements versus budget, time and quality. Architects are great when it comes to designing houses based on a particular vision, but it is not their job to fully understand the financial aspects of the construction. That is where your job comes in!

Building Site

Whether or not you have space around your project, there are still things you will have to consider that concern the actual building site. The less space you have, the more planning is needed. When material will be delivered, and there will be a lot of deliveries, the material might need to be fitted/installed quickly if no space is given. But let's assume you have got space on your plot to spare for your project, then you should create a draft of the space and sketch how and what things need to be laid out on the construction site. First make a note of the things you need to consider:

- Property boundary
- The project itself/the house
- Construction hut
- Skip
- Container
- Perhaps the caravan you plan to live in during the build
- Car park for a few cars
- Area where material can be dropped of
- A spot where a crane can be placed and mark out its radius

I would suggest talking to the contractor about this in order to seek a buy-in but also seek advice. Have a look at the example below:

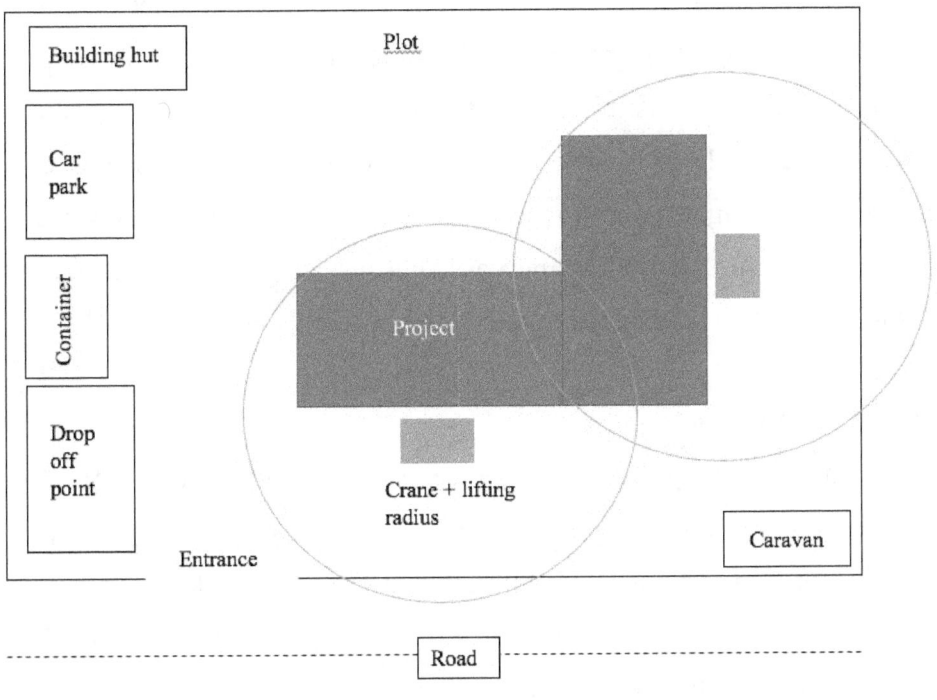

A building hut with toilet might seem like an unwanted cost. But it is most convenient to have a place where you and the workers can sit and plan, eat during the day, change cloth or even be able to take a shower. This is particularly useful during windy and rainy days. Also use the hut walls to stick various pieces of information to, like the time schedule, a to-do list, contact information, a picture of the end goal but also a place to keep drawings and other useful information.
Of

Furthermore, keep a first aid kit on site. Having a container on site will also be useful for storage of building material and equipment such as ladders, plastic covers, tarpaulins and so on during the project. But if there is little space there will be other decisions to make. You might need to use the road outside or perhaps the neighbour's site. If that is the case, permissions might be needed which could come at a cost.

Tidy Project

You should always ensure to keep the building site tidy. A suggestion is to spend 15-20 min at the end of each day tidying up, mainly for two reasons. You will have an efficient workplace where health and safety are safeguarded. Print out a few A4 signs around the workplace to help everyone remember!

Tools needed:
- Shovel
- Broom
- Floor squeegee (removes dust and water)
- Large bin bags easily accessible
- Large wheelie bins (enables you to go from room to room efficiently)

When the project is getting finished you will obviously have to use a vacuum cleaner, mops, mop bucket, detergents to get the place spotless, but not during production. So, plan for this activity as well.

When you are approaching the end of the project, or at the point where the floor is in place you might want to supply protective cover, like heavy weight paper or Masonite because hand trucks will most likely be used when assembling the kitchen or parts for the bathroom. If the staircase is delivered at an early stage of production, the necessity for floor protection is obvious.

Temporary Solution

Once the roof and exterior walls are up and depending on the climate, the openings where windows and doors will be should be covered. Covering all openings with temporary solutions might be worthwhile and arranging temporary heating or air conditioning to create a good working environment will lead to a more efficient production.

Cover the openings with tarp or plastic sheeting and get hold of an old door, preferably a bigger one so all the material that goes into the house will go in easily. It is all about making the work situation as efficient as possible all the time in a smart way.

Temporary Building Actions

In order to deal with a lot of heavy rain which will most likely create a very messy building site in terms of mud, it might be worth creating pathways using either wood or cheap gravel. It might pay in the long run since the need for cleaning will be less, and you will not be dragging damp into the house to the same extent while the project is progressing.

Underestimating temporary solutions will cause unnecessary stress and increased cost to your project.

Building Material Outside

Be aware of certain materials that require maintenance or storage when exposed to various weather elements. A wooden decking e g is one of the materials that will be exposed to rain, wind, heat and perhaps snow. Ensure that the wood is already well impregnated to withstand the weather elements. The frame that lies underneath the actual deck should not be laid directly on the soil as it attracts damp. Lay the deck frame on brick or stone for instance.

If you use tiles outside, you want to make sure that the grouting between the tiles is done correctly so it does not crack. Be aware of the slippery surface tiles have during rain or winter. Once again, I recommend you talk to the seller about the work that needs doing, so they can advise you the best way.

If you desire to use wood panel on the facade, make sure the contractor uses screws that do not corrode. The same goes for wooden decking. Screws that rust will have a devasting effect with brown stains all over the facade. And ask for references, preferably about a project that might have used the specific product 2-3 years ago so you can visit and judge for yourself what the end result looks like after being exposed to all weather seasons of the year.

Your contractor is hopefully familiar with the tips mentioned but it is not a guarantee. You will have to ensure, while the work is being done, that the chosen material will be sustainable.

Foundation of the Building

There are so many ways for the activity foundation to go wrong depending on the location of the site and how suitable the ground is. I cannot stress the level of planning enough when dealing with the foundation of a house build So, how do you go about this?

Bring in:

- Geotechnical engineer specialized in geological surveys. They will do the necessary ground tests where the building will be. This usually involves drilling in a few spots until they hit bedrock, and from those findings the necessary recommendations of how to build a solid foundation will be made.
- Structural engineer will calculate if the foundation can hold the pressure of the building, and additionally give advice on any other matter that might be useful.

By doing this, you will remove a great deal of risk from your project and at the same time help the foundation contractor produce a reliable quote for the slab and the underlying ground for instance. If the structural engineer recommends the use of piles, then a pile sub-contractor will be hired. There might be some uncertainty left after the geological survey, but hopefully around 80-90% is covered, leaving only a minor risk which might cause an extra cost. To get a more accurate foundation budget, you need to discuss the additional cost with the contractor if big boulders (found) need to be removed from the ground and what preferable method to use.

By doing this, in the event of finding big boulders you know what it will cost, being a known uncertainty so to speak. If the geological survey suggests a few piles for the foundation, find out the additional cost for one or two more piles once the production has started. Talk to the contractor about it. With the known uncertainty left, move it to your risk list and set a budget for it.

Here's an example how to work with an activity regarding risk. Let's say the bid from the contractor is €100,000 for constructing the slab. The contractor declares it is with a 90% certainty. The 10% uncertainty equals €10,000 for risk. You can make an agreement with the contractor that he or she will get €10,000 if the contractor encounters situations (risk) that require removing boulders or adding piles, but no more than €10,000. It is important to get as close as possible to a fixed price when it comes to the foundation. And the only way to achieve this is by hiring a geotechnical engineer and a structural engineer, since they will be able to produce reliable information on the foundation to the contractor who needs to feel safe when making a bid.

If you decide on underfloor heating or other installations in the slab it should be clear where in the slab they should be positioned, because once the concrete has hardened, things can be very costly to change. Below are a few common installations:

- Inlet of water pipes
- Plumbing
- Electricity
- Fibre
- Perhaps some metal columns require a metal plate to be submerged in the foundation while the slab is hardening

If the contractor does a good job with the slab, you do not have to order putty to level the floor which will save you a few thousand. Do not take for granted that you will get a 100% level concrete slab, ensure to reach an agreement about this. If possible, share this risk with the contractor. If the contractor fails to lay a 100% level slab, then pay 50% each of the cost that the putty will bring. This information is an example that should be stated in the SOW which will be part of the agreement.

Disability

If you or a loved one has a disability or perhaps you just want to plan for a later stage in life, it would be worth hiring a consultant, whose expertise on understanding the needs of a disabled person would be most valuable. For a person in a wheelchair the basic things are door width, low or no thresholds and lower height for appliances and countertops in the kitchen, bathroom, bedroom and so on.

Depending on the disability at hand, one option to fully understand the other individual's needs is to put yourself in their shoes. Try manoeuvring in a wheelchair and write down everything you struggle with. When deciding on the interior it is important to fully understand the needs. Two floors might even require a mini lift (big cost).

Windows

When choosing a window, a function called Tilt and Turn or called Kipp/Dreh by others, allows the window to open in several ways. It is a good design if you want fresh air without having to open the window completely or being afraid of a toddler falling out while getting a breeze.

Apart from this, consider the window frame that faces inwards in terms of the aesthetics.

As discussed under environmental aspects, triple glazing has some advantages:

- Keeps the heat inside better, energy saving +
- Good acoustics qualities (keeping noise out) +
- Keeps cold out +
- Higher initial cost -

The Sun

It is important to consider how the sun will shine on your house, since it will affect your living. If you wish to have big panoramic views for instance, you will likely experience a hot room. Therefore, you will have to make sure you have the right ventilation in order create a comfortable space. Windows also provide views and natural light in the room but will also transport cold. These things need to be considered depending on what you want to achieve with your house and what the weather will be like where you want to build.

For future comfort you might want to consider fitting awnings, blinds, or drapes or perhaps ducting for the purpose of having electrical awnings. Planning a ventilation system is also important to create a comfortable living space. This can of course be installed after the windows but bear this in mind when calculating the project.

Windowsill

When it comes to the window ledge there are two aspects to consider. Strive for a design that is timeless and works with any curtains and is durable. Preferably some kind of stone material, as you do not have to worry about marks when watering your windowsill flowerpots.

Acoustics

Acoustics is a subject that usually no one pays much attention to when it comes to building a new home. But sound plays an important role how you feel, (cause echoing noise will occur due to the lack of absorbing material) once the project is completed. Hard surfaces allow sound waves to bounce around.

If you plan to have a large living room, for instance, or any other big room for that matter, it will be necessary to install acoustics panels in the ceiling which is a good way to absorb sound. Panels are questionable when it comes to aesthetics, there are, however, hidden panels that could be fitted.

Furniture, curtains, carpets and any soft fabric are effective means to create a comfortable acoustics atmosphere in the room. But if acoustics are important to you, you should talk to the contractor about it or find a solution in the planning phase. Or try to find information about this in hardware stores.

There is also a great variety of walls and doors between rooms on the market, so your choice of wall material or choice of door, will affect the acoustics since they all have different characteristics. Doors that can withstand a great deal of noise are usually very heavy, so bear that in mind once they are installed. Make sure enough manpower is present and the right tools available when mounting the doors.

If you plan to install an advanced or expensive Hi-fi system, you could probably get some help from the store where you buy the system. A service minded seller will provide useful information when it comes to installation and acoustics.

If you desire to install tiles in your home, which could be doable if you live in a warm place or if you installed heating in the floor, make sure to look into a sound proofing system for that too. It usually involves placing a hard mat which is 1 inch thick between the concrete and tiles.

HI-FI System

You might also want a surround system for your television. Here it would be good, to speak to a representative at a store, helping you point out where on the drawing of the living room you should install the sound system and at what height the speakers should be placed. Other points to decide are what material the speakers should be mounted on. Decisions on visible or hidden wiring must be made because the latter requires ducting in the wall.

Fire Precautions

Do not forget to install 1-2 fire detectors in your home. They do not cost much and are a life saver in an unfortunate event. They do not need wiring as they run on batteries. Fire extinguishers are good to have, but costly and the life span is around 10 years and not very aesthetic to look at.

Motion Detector

Motion detectors which activate light can be useful for many purposes. Not only will they be convenient when you get home at night but will also deter unwanted visitors. The cost of installing this function is low so it is well worth looking into.

Characteristics of the walls inside the house

Once you have decided on which wall you want the TV placed or where you wish to place a wall-hung bookshelf ensure to mark it on the drawing. The wall requires a wall being able to hold the weight over a long period of time. A plaster board will not be enough. It needs to be supported preferably by a wooden board behind it. Make sure you provide the contractor with this important information.

Garage

A garage usually serves many purposes. Apart from the main purpose i e to keep your car, the garage is often used for storage and is combined with a utility room or even a workstation for fixing bikes and small carpenter jobs, etc. Remember this when you plan your garage together with the architect, ensure you think of the extra space needed for the things you wish to have.

To invest in a garage door operated with a remote control is a convenient way to reduce the problem with garage doors that get stuck due to ice and snow in the winter.

Consider this when choosing a garage door for your project. It can be resolved with salt or some heating in the concrete just where the garage door touches the floor.

Wet rooms

A sauna, bathroom or a utility room with dryer and washing machine will always generate moist. To deal with this and remove the risk of mould, fans or ventilation should be installed for the purpose of removing damp from the room and lead it out to the open.

Storage

When discussing storage, really discuss the needs your family have. Ask yourself the right questions. If you plan to have a garden, consider where you will store the garden tools and where you are going to store ski equipment, winter clothing and so on. Also discuss what kind of storage, wardrobes or a dedicated room just for storage.

Something that is becoming more popular in modern kitchens nowadays is a pantry, which is a great place to keep food that has a long-life span.

Fireplace

A fireplace can be very pleasant during autumn, spring or winter. Some models can give a significant amount of heat which will contribute to keeping your house warm. Depending where you live, the installation of a fireplace may require a signed form by a fire surveyor who can certify that everything complies with all safety rules. Remember to plan space for firewood.

Security

When it comes to security in terms of locks for your front door or any other door leading out to the open, the insurance company might give favourable quotes on their insurance as the future home will be more secure.

Furthermore, you might want CCTV or doorbells equipped with video function which require wiring or sockets, depending on what kind of system you choose. Make sure you place the sockets where you want the CCTV placed and clearly inform your sub-contractor who will do the electrical work about this arrangement.

Toilet

For maintenance purposes, a producer like Gerbait offers a sleek solution where the toilet hangs from the wall rather than being fitted on the floor. This will make floor cleaning so much easier. The flush buttons are fixed onto the wall, too. This solution requires installation done within the wall on a strong wooden frame which can withstand a great deal of weight. Make sure the plumber has installed one of these before, even if Gerbait can provide the necessary installation manual.

Garden

Do not forget, that you might also need a budget for a fence or a hedge round your garden/plot. This is also a point where you need to think in terms of maintenance. Do you see yourself trimming the hedge now and then, or do you rather strive for a solution that has low maintenance i.e. a brick wall or metal fence (painted or galvanised).

Driveway

The driveway will have to be functional in all kinds of weather and cater for the kind of cars you plan to have in the future. If the driveway is at a steep angle and you cannot do anything about it, it might be worth having a partially heated driveway so you can use the driveway during winter. Furthermore, make sure you can get in and out of the car easily, particularly when carrying shopping, tools etc. All this requires some planning, of course, and the installation of heating obviously comes at a cost. A project planner would have to look at this to find a cost-effective solution. Preferably find a company that will install it at a fixed price. Once again it will be you who will decide if it will be worth it or not.

Miscellaneous

There will be "things", small items that you will need for your house. For this, I suggest you just put miscellaneous in your budget and set a sum of € 2,000. It should cover things like:

- House number
- Mailbox
- Equipment for the garage such as workbench, tools etc
- Equipment for your garden such as an electrical lawn mower, hose, tools
- Fire extinguisher or fire detectors

Preparations for Future Needs

Electric vehicle charger:

A possible future electric car will require a vehicle charger at home, so plan for its position from the very start. Perhaps ducting is needed from where the charger will be placed and lead all the way to the main circuit breaker. You need to ensure that the main circuit breaker can handle more electricity in the future. At the same time, you might have thought of installing solar panels which can provide electricity for your home in the future. Here comes the concept again, correlations between activities. All this preparing requires that you consider the importance of ensuring that the electrical wiring to your house and the electrical circuit can handle future electricity. Make sure to take this into account.

SWOT Analysis

With the SWOT analysis, you will be able to understand important aspects related to your project in a more structured way. The analysis is widely used in the professional arena, especially within project management. SWOT will also help you during the planning phase, when you want to understand various aspects of materials and production methods. It will aid you when deciding how to proceed!

An example how to use the SWOT for the Project:

Strengths	Weaknesses
• Able to live close to the project • Live in a caravan	• First time building a house • Strained budget
Opportunities	**Threats**
• Grandparents able to help 3 days a week • 20% Discount at Hardware store • Borrow scaffolding from a cousin	• Plenty of rain during construction • Wind • Delays

SWOT Actions

Once you have gone through the Swot analysis, you will have to consider how to respond to the specific topic. To get help with this, you can either speak to the contractor or discuss your concerns with a representative at the hardware store.

Action steps **Rain:**
- Keep cover handy to protect material when it is raining
- Create friction on slippery surfaces
- Have boards or gravel handy to walk on instead of mud
- Keep a pump handy in order to remove water (if building has a basement)
- Look into the production method of prefabricated exterior walls and roof

Action steps **Wind:**
- Plan well in advance when acquiring a lift for the project and what time during the day will be most favourable to do the lift
- Hire/buy some heavy-duty support tools for the walls until the roof is in place

Action steps **Delays:**
- Plan accordingly, ask people involved questions. Try to work with standardized products (easy to get from the shelf). Work with reliable people. Mentally embrace that things might take longer than initially planned.

Action steps **Strained Budget:**
- Really go through the WBS together with the contractor(s), friends and family in order to expose activities that have not been accounted for. Then attach the cost and the amount of time to complete the activity.

Opportunities

Here, I really want you to think. Turn every stone possible in order to find possible opportunities that might be beneficial for your project.

A few examples:

- You will need to rent a crane during the project in order to lift and fit large windows. This could also be a good opportunity to use the crane for several other things, like lifting a huge number of plaster boards or insulation into the house. To do this you might have to leave a gap in the roof, which temporarily needs to be covered to keep downfall out of the building.

- If space is very tight on the building site, you might be able to use or hire some land from your neighbour or from the local council. Having some extra space is an advantage, as it gives you and the contractor flexibility when planning work that needs doing. Due to the fact you will not be constrained as you can take deliveries whenever for instance.

- The vision of your project might involve costly solutions at first. Work with the contractor and architect during the planning phase to find a solution that is both appealing to the architectural design and cost effective. There are always several ways of carrying out an activity, just make sure you can live with it! Use the experience of others to find a solution that will work for your project.

- Let's say you have a long wall/facade of 10-15 metres where you have envisioned a row of glass windows only to realize it will be too expensive. Perhaps you can come to terms with that fact, and instead decide on installing only 6 metres of glass windows and the remaining part will be a low cost wall, which can easily be altered once your finances will allow that, maybe 3-5 years after the project is finished.

- You have envisioned the idea of having solar panels on the roof or being able to charge your electric car on the driveway. You can always prepare ducting for these small actions. You might even prepare the roof for installation of solar panels. The same goes for preparing for lighting bollards in a garden or a small shed.
- Perhaps there is an opportunity to use standardized products, lengths of products that are easy to get hold of or easily produced without compromising the architectural design.
- Depending on what design you like and prefer, you might wish to have electric wiring and pipes for the heating system on the outside of the wall. This will greatly reduce time and effort in terms of planning.
- If you have plans to build a garage, you might want to do that first and then use the garage as storage during production, or even a place to temporarily stay during production depending on what your final intentions are with the garage.

Obviously, most of us want a 100% completed project to move into. But only you and your partner can decide what 100% is for you at the given time.

Risks

This is one of the most important issues, as they will make you aware of risks associated with the project and the actions you and your partner will choose when proceeding with the project. Ask friends and family to participate, in order to bring an unbiased review of your decisions when identifying risks. By planning, investigating and clarifying things, the less risk, you, contractors and sub-contractors will take. Quotes will consequently be more reliable, since there are no uncertainties that will not be priced.

It is somewhat unconventional to view family as an area of risk. Understandably it does not come across particularly charming. But it is a fact, how are you going to ensure to keep your family intact while the project is going on? Take the question seriously.

Below is an example of to use a risk analysis:

Activity	Consequence		
	Mild (1)	**Harmful (3)**	**Serious (5)**
Improbable (1)	Insignificant risk (1)	Acceptable risk (3)	Moderate risk (5)
Possible (3)	Acceptable risk (3)	Moderate risk (9)	Significant risk (15)
Likely (5)	Moderate Risk (5)	Significant risk (15)	Unacceptable risk (25)

Actions

As the risk is considered insignificant (1) no actions are needed.

Acceptable risk (3) no actions are needed but should be reviewed and controlled until they are removed.

When entering moderate risk (5 or 9) preventive actions should be taken.

Significant and unacceptable risk (15 or 25) must be fixed before proceeding. Really make sure these risks are thoroughly thought through and dealt with.

Risk Assessment:

Area of risk	Explanation	P	C	R	Action
The project					
Foundation	Boulders are found when creating the foundation	3	5	15	Set aside €10,000 in risk reserve – get the buy in from the engineer and contractor it will be enough
Continuing rain & wind	Delays work	3	5	15	Set aside €5,000 in risk reserve– get the buy in from the engineer and contractor it will be enough
Building site	Extra pump, extra temporary material needed	3	3	9	Set aside €2,000 in risk reserve
Hire of crane	you need to hire a crane more than two days	3	5	15	Set aside €2,000 in risk reserve for two extra days

Theft	Material can get stolen	3	3	9	Set aside €3,000 as risk reserve
The Family					
You lose your job	Very unlikely since you have a very stable employer	1	5	5	You have insurance that will cover 80% off our wage if you for some reason lose your job
Partner	Level of stress	x	x	x	Make a sustainable budget, make a reliable time schedule. Do date-nights, help at home
Son1	Not performing in school	x	x	x	Buy help for private tutoring
Daughter 1	Feels neglected	x	x	x	Make sure there is help, Grandparents will help. Give your partner money so he/she can have more fun with friends. Be there at least twice a week to help and socialize.

The document assessment of risk should be revisited continuously as you go along with your planning, to remove risk, add risk, reassess risk.

Budget

The calculation is based on quotes from contractors, sub-contractors, prices from warehouses, local council and so on. The contractors and sub-contractors have based their quotes on the information you have provided them with regarding materials, production method, quality and time schedule. Once you are done the calculations will be your set budget! The budget should be 96-98% correct if not 100% at this point closely corresponding to your estimates, if not, you need to go through all the activities again together with the risk analysis. What have you missed? What estimate is too unsure? Are all the assumptions and expectations clear with the potential contractor and sub-contractor, i.e. do you speak the same language about the conditions regarding the project? What activities have you deliberately not priced, if so, why? Has the budget been scrutinized by friends and family?

The budget displayed below has not considered the costs that come with temporary living and other costs that have to do with expenses that you and your family have besides the housing project. Create the budget specifically for your project based on the direct expenses that come with it, not indirect costs.

Estimate activity cost = Is the cost that will vary during the planning stage depending on what material you use and based on the quotes from contractors. Every material and production method has its own estimate activity cost, so when you plan and look at all the different choices the estimates will vary in your calculations.

Budget = Before you start production.
Final cost = The actual cost of the finished project.

During production you should continuously revisit your budget and go through the actual cost. It is called prognosis. Did something after all cost less or more? What do you need to do, in order to get on track with the budget? Did you miss an activity despite all the hard work with the calculations? You can easily make a budget and prognosis on a Microsoft excel sheet together with the payment plan provided at the end of the book.

Activities	Estimate costs	Budget	Final
Plot	100 000	100 000	
Permission	2 000	2 000	
Taxes	30 000	30 000	
Water main	7 000	7 000	
Sewer	8 000	8 000	
Electricity	5 000	5 000	
Fibre/Copper Internet	3 000	3 000	
Building site	130 000	130 000	
Architect	4 000	4 000	
Geological survey	5 000	5 000	
Structural engineer	2 000	2 000	
Interior designer	1 000	1 000	
Foundation	30 000	30 000	
Steel frame	4 000	4 000	
Exterior walls	30 000	30 000	
Windows	40 000	40 000	

Interior walls	30 000	30 000	
Inner doors	15 000	15 000	
Kitchen & appliances	30 000	30 000	
Flooring	12 000	12 000	
Tiling	50 000	50 000	
Ventilation	6 000	6 000	
Electrical	14 000	14 000	
Plumbing	50 000	50 000	
Painting	50 000	50 000	
Crane	30 000	30 000	
Furniture	6 000	6 000	
Hi-Fi system	2 000	2 000	
Landscape	5 000	5 000	
Estimate cost	701 000	701 000	
Risk reserve	22 000	22 000	
Contingency reserve 5%	55 000	55 000	
Total:	778 000	778 000	

Time Schedule

Let's discuss the time schedule! Using the "tool" WBS when planning you will get a clear picture of all the activities that need to be done in order to complete your project. Just like the estimate activity cost, the WBS changes during planning until you are ready to create and set your budget. It is understandable that not all activities can be done at once. You cannot paint a ceiling before the ceiling is in place, right? The time schedule is there to give you an overview of the whole project and enable you to plan when the specific activities should take place. When working with the time schedule, a couple of things are most important:

- Speak to a couple of contractors about the time frame needed for each activity. Let's say the painter needs 3 weeks, the contractor needs 5 weeks to do the slab and another 5 weeks to mount steel frames for the exterior walls. During this work, do not take the time frame they give you for granted, discuss their assumptions and expectations with each contractor once he or she gives you the estimated time frame.
- Understand if an activity is correlated with another activity. Here is an example of jobs to be done when fitting the exterior walls. The electrician needs to do the ducting in the walls in order to "hide" all the wiring, or the plumber needs to extend some pipes from the slab within the exterior wall. It is therefore most important to understand which activity will affect another, so ask the right questions when creating the WBS and make the necessary notes!

Depending on how comfortable you are with the estimates for each activity, you can always add a few days in between, in case the specific contractor needs a few more days to complete the work. That way the contractor will not interfere with the next contractor.

Below is an extract and example of a time schedule that you can easily create in Microsoft excel. Here you can see activities from the WBS showing activities that intervene with each other and what activity that follows another. The red stands for three weeks over holiday. Numbers represent the weeks which are the estimate given from the contractor including a few extra days.

Project	Weeks	jul				aug				sep				okt				nov				dec				jan			
		27	28	29	30	31	32	33	34	35	36	37	38	39	40	41	42	43	44	45	46	47	48	49	50	51	52	53	1	2	3	4
Production																																
Foundation	8																															
Steel frame	5																															
Exterior walls	4																															
Roof	7																															
Windows	3																															
Interior walls	7																															
Plumbing	4																															
Electical wiring	4																															
Ventilation	4																															
Painting	7																															
Flooring	8																															
Kitchen	5																															
Bathroom	5																															
Garden	10																															

When acting as project manager do not just assume that everyone will turn up on the expected date shown in the time schedule. Call the specific contractor to find out if there is any lead time and ask for and any information that might jeopardize the time schedule at all. You need to be on top of this!

Lead time is from when an order is made of a particular product until its actual delivery. If you order a "unique" product, lead time might for example be 10 weeks. If a manufacturer says 10 weeks, then you need to ask what the likelihood is that the delivery will be in 10 weeks.

Communicate with the manufacturer during the lead time to find out if the delivery time is still intact, do not just wait until week 10. You need more solid information to go on once you book crane and manpower.

You can also use the time schedule for your family activities, such as holidays, date nights and sporting events for your children. The importance of keeping family life functioning despite the project cannot be emphasized enough. The "whole picture" must be considered to make the project successful.

Note all the planning activities that need to be done before the production phase. Point being all planning should be finished before production that is the general principle. However, if you intend to hire a contractor finishing the landscaping or the kitchen, there will be time to wait with those procurements as those activities are so far in the production. You do, however, need to set a realistic budget for those activities.

Planning permission and being able to finance the project are the critical milestones that will have to be in place before proceeding.

2019-06-23		2020																			
		mar				apr				maj				jun							
Project	Weeks	8	9	10	11	12	13	14	15	16	17	18	19	20	21	22	23	24	25	26	27	28
Planning	36																						
Architect	9																						
Geotechnical engineer	9																						
Structural engineer	9																						
Agreement documentation	17																						
Swot analysis	17																						
Risk analysis	17																						
Time table	17																						
Build site planning	17																						
Planning permission	10																	■					
Budget																							
Loan from bank																							

Use Microsoft excel preferably and start creating a time schedule. Excel has even got finished templates that you can use for your project.

Notes – how long will it take to create the foundation, exterior walls, roof? Talk to the contractors.

Planning Done

Congratulations, you have done a great job! By this point you have a good idea of what your project will cost, when it will be finished (date for moving in) and what quality to expect. Your partner is hopefully also satisfied with all the provided information. So now you are ready to sign a contract with a contractor and provide a SOW and other agreement documents required, so everyone knows what each party should do and when. We will get to the agreement shortly.

It is always good to go back and go through all the parts of the project and have friends and family to go through the drawings and have them ask questions one more time. By now you should have all the answers anyway, but if you do not, it might be worth investigating the issues further.

Remember, you should be crystal clear what is going to be built. You need to visualize from the drawings what the end goal will look like. If that is difficult, have the architect produce 3D drawings to further help you and your partner to understand what your build will look like in detail. You do not want to be the project manager who changes things during mid-production. Such changes will <u>cost and delay the build to an unknown extent, due to the fact that</u> the contractor has already bought the material ordered in the first place, hired a certain amount of tradesmen and new drawings, new material will be needed and so on. This is trouble you want to avoid as much as possible.

The Agreement

Depending on what country you live in, there might be various kinds of regulated contracts in a particular shape or form. However, the presented approach in this chapter will describe a universal approach to reach an agreement. The contract should be legally binding between two parties and should define scope, cost, time schedule and quality which both parties agree to.

The Agreement with the Contractor

I will only address two approaches when it comes to an agreement when hiring a contractor. Fixed price, time schedule and materials.

Fixed Price

With a fixed price or lump sum, you know exactly what the job will cost. You do, however, need to produce an agreement that describes in detail what the work entails. The document "statement of work" (SOW) does exactly that. The agreement should be supported with drawings in order to bring as much clarity as possible. Let's say, you want to hire a tiler to do the bathroom, then it would be very beneficial to attach a drawing of each wall and floor with the pattern you desire. Together with the statement of work (SOW) the agreement can state things like:

- Putty is included. Angle from door to drain should be 1cm per metre. (All water will run to the drain)

- Waterproof membrane should be put underneath the tiles (product and ID)
- Tiles (product and ID) are included in the agreement
- Installation of drain is included
- Making all the holes in walls for electricity and plumbing is included
- Payment should be paid in full, once the whole job is completed
- The client (you) is obliged to produce water, electricity, bathroom and some lockers
- The client will provide a parking lot
- Painting of the ceiling and spotlights are NOT included.
- Under floor heating is NOT included in the agreement.

The best way is to discuss what, when and how things should be done in order to avoid false assumptions and thereby align your and the contractor's vision of what is going to be built and under what circumstances. You want to be at ease in your collaboration and the same goes for the contractor. Regarding under floor heating or ventilation, it is important that the contractor is accountable for the function. The installed product must function just the way it states in the product specifications.

Time and Materials

When you hire a contractor based on a time and material contract, you will know exactly how much the contractor will charge per hour. Furthermore, any percentage the contractor has added for purchased material according to the SOW or room description specifications, will allow you to know exactly how much you will have to pay. Before entering this kind of agreement, a budget should be discussed so you have an idea how much it will cost in the end. The following questions needs to be answered:

How much manpower will be needed?
How long it will take?
Do you have to pay any percentage on top of the contractor's purchase price of materials?
Can you strike a deal with the contractor where you can benefit from any discounts they may have at the hardware store, or will it be cheaper if you deliver the materials to finish the job?

Statement of Work (SOW)

The SOW is a long list of things you want to be included by the contractor, but you can also put in assumptions or expectations in order to be as clear as possible.

Depending on what sub-contractor you want to hire, you provide the procurement documents for the agreement in question.

Room description

Once everything is decided what you want in each room regarding floor, walls and ceiling you should produce a document (preferably A4) called room description. Let's see what one page could look like:

Living room:

Floor	Hardwood oak (product no & ID)*
Plinth	Follows the same colour as the floor (oak)
Walls	Wall paper (product & ID no)*
Ceiling	White (product & ID no)*
Cornice	-
Interior	-
Other items	Shelving unit (product & ID no)*

You have visited a store beforehand and chosen what products you want, and with that information, you will be able to complete the document room description.

Maybe it looks a bit dull. Let just say, you want to paint the walls like a basketball court, then just write it down right beside "walls" and what paints you need. Once again, you have talked to a painter about it and got a quote for it.

Once you have done all the rooms in the project, you will have achieved a great overview and added clarity to what each room will eventually look like. The contractor will also see exactly what you want and what is included in the specific agreement. The document room description will also work as a great communication tool for everyone involved in the project, so make sure you create a document with each room described.

- Living room
- Kitchen
- Bathroom 1
- Bathroom 2
- Bedroom 1
- Bedroom 2
- Bedroom
- Spare room
- Entrance hall

You should also be able to see each room clearly on the architectural drawing.

Specifications and Measurement

The architect will be able to help you by providing a simple table sheet of how many doors, what sizes and information whether the door should be left or right hung. The same goes for the windows needed. Measure and note the floor and wall square metres on each room drawing. This information should be used when ordering the specific product or used when reaching an agreement with the specific contractor/manufacturer.

Example of a Contract

Here I will give some examples of what an agreement could look like with a particular contractor for a particular job. You adapt the contract documents to the specific agreement. Let's look at an example for the kitchen and see if it makes sense to you:

1. The contract

a) states the amount to pay for the work and when

b) how long the work will take

c) what documents account for the contract

d) signatures and date

2. Drawings of what the kitchen will look like and:

a) Product & ID of faucet and sink (plumber)

b) Led lamps that might have to be installed in the kitchen (electrician)

c) Product & ID of tiles (tiler)

3. Room description

4. SOW

a) The contractor will buy, deliver, assemble and fit the kitchen according to the drawing dated 2021-xx-xx
b) The contractor will install the appliances (product list & ID)
c) The contractor will hire and be responsible for coordination with electrician, tiler, plumber and painter.
d) The contractor must cover all vertical surfaces with Masonite for protection. The client and contractor will check the finish together.
e) The client is responsible for providing a trash bin.

Perhaps some of the points here are taken for granted by the contractor where you live. But if that is the case, initiate a discussion to really find out what is included or not in the job. Once again, it is all about aligning your and the contractor's assumptions and expectations to bring clarity to the job at hand.

The Contract

The contract itself can be viewed as a cover page where you will find information such as:

- The total sum, date, signature, quality specifications, regulations of financial character, arrangements regarding payment etc.
- Documents related to the contract such as SOW, drawings, drafts, room description, time schedule and reports.

To-do List

Once you have established all the activities regarding foundation, steel frame, kitchen etc, then you can decide on a rough estimate where and when you should conduct the procurement and when you need the agreement finished. Let's look at how to conduct this:

You have made your time schedule and you see that the assembling and fitting of the kitchen should start week 35. You know it takes 8 weeks to manufacture the kitchen from the date you place the order and delivery takes one week. Consequently, you need to place the order no later than week 26, which also means that you should give yourself about 4 weeks to plan what the kitchen should look like. This will require that the drawings from the architect are finished so you know exactly where and how the kitchen will be fitted. Create a to-do list which correlates with the activities in the time schedule to help you take the right actions at the right time!

Safety

Safety is about leadership and attitude on large construction sites. Having said that, you won't get you off the hook. Do not take safety for granted. Usually stress and low safety budgets are the underlying factors why people are getting hurt in construction. People who are not accustomed to construction are not always fully aware of what could go wrong.

If you are going to participate in some of the labour yourself, ensure to use good ladders such as step stools.

It is of utmost importance to have a solid foundation to stand on! Do not take stupid risks, it is not worth it. If you are going to work on the roof, make sure using a safety harness unless there is a scaffolding around the project, and make sure windows on the roof are covered until the they are actually fitted. Make sure the workers in your project are aware of the safety aspects and respect them.

Be aware and use your imagination and try to foresee things that could cause accidents, like falling objects, slippery surfaces, heavy lifting etc.

The personal safety gear that is usually worn on a construction site consists of:
- Gloves
- Hardhat
- Protection glasses
- Vest
- Work pants
- Work shoes
- Ear protection
- Face mask

Depending on your role in the project, you should make sure you have the necessary safety equipment. You should also ensure that all of those working at your project write down their next of kin in case of an emergency.

Production

Once production finally starts, you should all the time you spent planning bear fruit. The contractors will produce according to the timetable, deliver the expected quality and produce what is stipulated in the contract agreed upon. This will reduce possible stress for your family.

It's during production you will act as the spider in the web, calling the various contractors, dealing with any building implications, overseeing the time schedule and budget and making necessary adjustments. Beside all that, you need to take care of your family.

All planning permissions from the local council or equivalent should be in order before proceeding into production. It is crucial that you follow any rules or laws that apply in your area/country during production.

Be aware of making any small changes adjustments in production. If a contractor or you have an idea to change something, really consider what possible implications the change might have for any other working group on the project.

Independent Surveyor

Depending on where you live, it might be mandatory to hire an independent surveyor for your project. It is always good, particularly to ensure the framework is stable, electrical wiring is done correctly as well as plumbing, ventilation and any other installations. Insurance companies are likely to require everything to be done by the book, if not, your homeowners insurance might not cover. Check!

Gather the contractors who have done a certain installation and do a final check that everything is right and works the way it should before moving in, like a "handover". Have a checklist ready! As the owner you should be at ease with how to operate all the various installations once you have moved in. Ensure it is agreed upon with the contractor, that they are required to provide the necessary information and guidance for the different installations.

There is nothing worse than moving into a house not knowing how all installations work and how they correlate with each other once they are in place. This part of the phase can give room for a contractor to blame another contractor for not delivering what is required for their product to fully function.

An example:

Let's say the installed automatic garage door requires 400v but the electrician only produced 220v making the door not fully functioning. All these things need to be sorted before moving in, so make room in the time schedule for a final check-up.

Maintenance

For future purpose, write down what paint codes you have been using when painting (product & ID), what kind of light bulbs have been used and what kind of specific product & ID does it have. This is something the contractors can and should help you with. It will be a lot easier to have this documented on 1-2 sheets of paper than you having to get hold of the contractor 2-5 years later when you might have scratched a wall and need the original type of paint.

Accommodation for Tenant

Maybe you consider letting a part of the house or a separate building on the site, which is a great idea of course, since an additional income will be achieved by this. Just make sure a solid contract is in place with rules that are important to you. Apart from that, ensure you know the previous history of the tenant, such as:

- Previous references
- Employment history
- Criminal convictions

Do your homework to minimize any potential loss and trouble letting your property!

To make things smooth, invest in an electronic door lock. Every time a tenant moves out you simply change the door code and you won't have to worry about potentially lost keys.

Summary

There will be situations impossible to predict and consequently additional costs and delays that you have not calculated with. That is just a fact! But hopefully this handbook has given you a reasonable number of tips of the trade on how to handle unpredictable events that will occur. Events that are small and manageable due to the extent of planning. Ultimately the advice in this book will hopefully provide you with knowledge and confidence to build the home of your dreams within the time frame and budget you set for yourself. To have a positive attitude to the project is excellent and perhaps some would say essential in order to pull the project off.

If you have an idea how long the project should take or how big a budget you have for the project, do not let naivety taint the challenge and your positive attitude.

Ask yourself how you came up with the time schedule and the budget and let others scrutinize them. This must be done in the beginning when planning, not in production!

By now, you should realize that breaking down the different parts of the project in manageable pieces through WBS is important.

Documents you should work with:
- Feasibility analysis
- Detailed budget

- Time schedule
- Contract
- Room description
- Statement of work (SOW)
- SWOT analysis
- To-do list
- Risk analysis
- WBS structure

I urge you to hire:

- Architect
- Structural engineer to determine the best foundation & Frame
- Technical, geological engineer - for the foundation

I am glad to offer you templates that will aid your work as a project manager. Please visit this address:

https://tinyurl.com/y2ckvnww

You will need Microsoft word & excel

Final Thoughts

It is understandable that building your dream home is something you and your partner have been contemplating a long period of time and you cannot wait to get started to finalize the envisioned project. That is great, you are going to need that ambition and determination to succeed. You also need pragmatic thinking! What you do not need is naivety and false expectations. This can potentially ruin your relationship with several parties in the project.

By thorough planning and involving the people you need to hire, your expectations will be realistic and fall into place. Furthermore, this will be communicated to your loved ones and can be used when planning everyday life that goes on during the actual project. Building your dream home is going to be a challenge for you, but a challenge that needs to be overcome and celebrated. Celebrate milestones, e g the foundation is 100% completed, the roofing is 100% finished. Remember, you are not alone in this! Ask for advice, speak to the contractors, ask frequent questions to fulfil your own need of reassurance. This obviously requires people skills when dealing with a lot of contractors. And if you feel you are in deep water, you can always get support from professional project managers.

Wordlist

Concept	Explanation
Product & ID	Refers to a specific e.g. paint, floor, characteristic and producers own ID number.
Putty	Also known as floor leveller. It is a loose compound which hardens after 24h usually and is used when you want to level a concrete surface. It is also used in bathrooms to ensure the water flow towards the drain. Tilers are usually experts in this.
WBS	Work breakdown structure
Heavy weight paper or Masonite.	Usually used when protecting floor and walls.
Tarp or plastic sheeting	Plastic cover, perfect when protecting the house from rain but also material outside from rain and sun.
buy-in	Get a person or company to agree upon a suggested solution.
Hard mat	It is a kind of mats you put between concrete and tiles in order to remove noise.
Lighting bollards	Lightning for the garden, see hardware store.
Well impregnated	Decking are usually impregnated so the wood can withstand sun and rain for several years.
Trusses and sections of tongued and grooved boards to the roof.	Sections of wood you put on the beams/frame of the roof. It is where you will put the slates or metal sheet.

Please visit the Facebook & Insagram page and share your ideas about the book, perhaps your self-build as well and the experiences you had while building your home! You will also get good inputs about project management!

https://www.facebook.com

www.Instagram.com

The Facebook page will hopefully act as a community where project managers can share their experiences when they are building their dream home!

Notes:

Notes:

Notes:

www.ingramcontent.com/pod-product-compliance
Lightning Source LLC
Chambersburg PA
CBHW070653220526
45466CB00001B/418